Analysing and attempting to connect the genetic and metabolic derangements underpinning a disorder which is linked to schizophrenia in Irish high density schizophrenia families.

See preview to check for availability of free access.

Keywords/phrases: Schizophrenia; Irish high density schizophrenia families; estrogen response elements; P5CS; IL3.

Author: John Neville. **Contact:** jwilljohnwill@gmail.com

Published: 22.03.15

Remedial corrections to referencing: 23.06.15

ISBN: 978-1-326-22333-5

First edition

Copyright: John Neville 22.03.2015

ISBN: 978-1-326-22333-5

Abstract

A familial disorder that causes psychosis is described. Associated with the disorder are distinct patterns of physical symptoms and conditions that may be used to assist diagnosis. The proband carries genetic markers that link the disorder to schizophrenia in Irish high density schizophrenia families. The genetic and metabolic derangements underpinning the disorder are analysed and potentially linked. Evidence is considered which supports a working hypothesis in which haplotype SNPs located at IL3 and/or ACSL6 at 5q31.1 result in or from the failure or partial failure of one or more ERE to activate a pathway leading from P5CS to proline.

List of abbreviations:

ACSL6 - Acyl-CoA Synthetase Long-Chain Family Member 6.
AD - Arginine Decarboxylase.
AGAT – L-Arginine:glycine amidinotransferase.
BDNF – Neurotrophin.
BDNF-TRKB – Neurotrophic Tyrosine Kinase, Receptor, Type 2.
BDNF-AS - BDNF Antisense RNA.
CACNA1I - The pore forming alpha subunit of the Cav3.3 T-type calcium channel.
cGMP – Cyclic Guanosine Monophosphate.
COMT - Catechol O-Methyltransferase.
CSF2RB - Colony-Stimulating Factor-2 Receptor, Beta, Low-Affinity.
DTNBP1 - Dystrobrevin Binding Protein.
ERE - Estrogen Response Elements
FD - Frontotemporal dementia.
GS – L-glutamate-5-semialdhehyde.

GEP – Granulin-epithelin Precursor.
GRN – Progranulin.
hMATE1 - Human Multidrug And Toxin Extrusion 1.
HRE – Hormonal Response Elements.
IL3 - Interleukin 3.
IL3R - IL3 Receptor.
IL3RA – IL3 Receptor Alpha.
IHDSF - Irish high density schizophrenia families.
iNOS – Nitric Oxide Synthase 1. -
NO – Nitric Oxide.
OD – Ornithine Decarboxylase.
PD – Parkinson's disease.
PKG - cGMP dependent protein kinase.
PMP22 - Peripheral Myelin Protein 22.
PRODH - Proline Dehydrogenase.
P5CA – Pyrroline-5-carboxylic Acid.
P5CR - Pyrroline-5-carboxylate Reductase.
P5CS - 1-Pyrroline-5-carboxylate Synthetase.
SMAD5 - SMAD Family Member 5.
SMS - Smith Magenis Syndrome.
SNAP25 - Synaptosomal-Associated Protein of 25 kDa.
SNP - Single Nucleotide Polymorphisms.
SLC25A48 - Solute Carrier Family 25, Member 48.
SPRY4 - Sprouty Homolog 4.
TAAR6 - Trace Amine Associate Receptor 6.
TGFBI - Transforming Growth Factor, Beta-Induced, 68kDa.
TP53 - Tumour Protein 53.
VMT1 - Vehicular Monoamine Transporter 1.

INTRODUCTION

The weight of evidence suggests that genetic factors may prove more significant than environmental factors in the aetiology of schizophrenia and that in most cases schizophrenia may result from a complex interplay involving multiple genes. However, in familial schizophrenia it may be more likely that there could be a single genetic cause or that there is less complexity involved.

This paper examines a case of familial psychosis with a view to providing some insight into the genetic and metabolic derangements that can be involved in schizophrenia.

The hypothesis (see figure 1)

The evidence supports the view that:

1. A pathway from glutamate via P5CS to GS to P5CA and on via P5CR to proline (the proline pathway) could be TP53 and estradiol up-regulated/cortisol down-regulated;
2. An opposing pathway from proline via PRODH to P5CA to GS to ornithine and on via OD to polyamines (the polyamine pathway) could be cortisol up-regulated/TP53 and estradiol down-regulated; and
3. These pathways may be over or under-expressed in some forms of schizophrenia.

A working hypothesis which could explain the proband's disorder is that SNPs located at IL3/ACSL6 at 5q31.1 which are significantly associated with schizophrenia in IHDSF and which are located in or around EREs result in or from the failure or partial failure of one or more of these EREs to activate P5CS/the proline pathway in response to estradiol.

In the proband's family this results in mild or late-onset symptoms/conditions consistent with P5CS/P5CR disorders. It also triggers compensatory TP53 activity to up-regulate the proline pathway and results in over-expression of the opposing polyamine pathway. Abnormal TP53 expression also impacts on PMP22 expression, causing symptoms of a 17p11.2 disorder, and could affect GRN expression.

If the proband has under-expression of his proline pathway as well as consequential over-expression of the opposing polyamine pathway, this could result in under-expression of an arginine/arginase/OD/polyamine pathway and over-expression of an arginine/iNOS/NO pathway. This may be expected to impact on agmatine metabolism and on the estrogen regulation of AGAT and thus his arginine to creatinine pathway, and could perhaps also lead to over-expression of cGMP.

MATERIALS AND METHODS

Detailed information on the symptoms and conditions found in the proband and his family members is located in the supplemental material.

The physical symptoms/conditions

P5CS is located at 10q24.3 whilst P5CR is located at 17q25.3. P5CS and P5CR disorders are recessively inherited disorders that encompass two types of cutis laxa and geroderma osteodysplastica.

PMP22 peripheral neuropathy with liability to pressure palsies is a dominantly inherited disorder which results from mutations of

PMP22 which is located at 17p11.2. SMS is a developmental disorder which results from de novo deletions, also at 17p11.2.

Mild or late-onset symptoms/conditions consistent with P5CR/P5CS disorders and 17p11.2 disorders are found in the proband and run in the proband's family on the paternal side in those family members who have suffered from delusional beliefs. The proband had lipid abnormalities consistent with the lipid abnormalities that can present in SMS; (1) the father had similar more severe lipid abnormalities.

The neuropsychiatric presentation

The proband is forty-five years old. In his thirties he began suffering debilitating insomnia and episodic psychosis accompanied by severe anxiety and panic. Onset of significant psychiatric symptoms was preceded by obsessiveness, night-terrors and sleep disturbance that commenced decades earlier. Memory problems occurred many years after his initial psychiatric symptoms.

Episodic psychosis is found in at least three generations of the proband's family on the paternal side. Memory problems and/or symptoms of PD often follow or precede the psychosis. The paternal grandmother suffered psychosis associated with dementia and PD. There is a pattern of obsessiveness, but not one of anxiety and panic, accompanying the psychosis in family members.

Female family members are often most affected by the disorder. Only 2 out of 12 of the last two generations are female (a mother and her daughter), which it is speculated might, in part, be because of miscarriages of female fetus. The proband reports that in him events associated with estrogen result in diarrhea and events associated with cortisol result in symptoms of psychosis.

Biochemical and genetic tests

Biochemical tests were primarily performed at Westmead Hospital Sydney, Royal Prince Alfred Hospital Sydney or The Hospital for Sick Children Toronto. SNPs were tested by a private genetics company.

RESULTS

Abnormal pathology

The proband's abnormal biochemical test results are provided in the supplemental material.

SNPs carried by the proband

Chen et al (2) reported that rs440970(A/C), rs31400(C/T) and rs2069803(C/T) which are located at ACSL6 to IL3 at 5q31.1 are significant markers of schizophrenia in IHDSF and that these SNP are located in and around ERE1,2,3 and 4. They stated that ERE are promoters and enhancers of the IL3 gene, though they highlighted that the regulatory targets of ERE1,2,3 and 4 have not been identified. They also reported that rs3846726(C/T) is another significant marker and that two and three marker haplotypes involving key SNPs (rs31400(C/T), rs31480(C/T) and rs2069803C/T) are also significant. The associations were largely derived from female subjects. There is also evidence to suggest a joint effect between rs31400 and the DTNBP1 SNP rs760761 (CC-CC,CC-CT,TT-TT). (3) The proband carries rs440970(GT), rs31400(TT), rs2069803(CT), rs3846726(TT) and rs760761(AG).

The following DTNBP1 (6p22.3) SNPs were reported to be associated with schizophrenia in IHDSF: rs1018381, rs2619522, rs760761, rs2619528, rs1011313, rs2619539, rs3213207, rs2619538. Of these rs2619539, rs3213207, rs2619538 were reported to be part of a protective haplotype (GGT) in IHDSF. (4) The proband carries rs1018381(AG), rs2619522(AC), rs760761(AG), rs2619528(CT), rs1011313(CC), rs2619539(GG), rs3213207(TT), rs2619538 (TT).

In a study of the COMT gene's (22q11.21) association with schizophrenia, also in IHDSF, Chen et al (5) reported that rs4680 had a modest association whilst haplotype analysis indicated that haplotype A-G-A for SNPs rs737865-rs4680-rs165599 was preferentially transmitted to affected subjects. In Ashkenazi Jews, rs737865 has been reported to have a significant association with schizophrenia with rs737865-rs165599 giving greatest significance, particularly the G-G-G or the G-G haplotype. (6) There is also a strong association between COMT SNP and Alzheimer's disease and psychosis in male subjects which appears to result from interaction of the ERE6/RS737865, RS4680, and RS165599 loci. (7) The proband carries rs4680(AG), rs737865(AG) and rs165599(AA). It is unknown if he carries the ERE6 SNP.

In a study on the neurogenin1 gene, again in IHDSF, Fanouse et al (8) noted that neurogenin 1 is located at 5q31.1 and that it is a transcription factor. They found the major alleles of rs8192558 and rs2344484 are over-transmitted to affected subjects and might have a small effect on susceptibility to schizophrenia. The proband carries rs8192558(AC) and rs2344484(GG).

Zahareiva et al (9) reported that Rs17169180, rs1859430, rs7715300 and rs6897690 are located in the chromosomal region 5q31.1-q32. Also, that rs7715300 is located between TGFBI and SMAD5 whilst rs6897690 is located in intron1 of *SPRY4*. They found significant association with schizophrenia at rs7715300(C/A) and at rs6897690(A/G) and suggestive evidence for association with schizophrenia at rs17169180(C/A), rs1859430(T/C) and rs7443175(T/C). The proband carries rs17169180(AC), rs1859430(GG), rs7715300(AA), rs7443175(CT) and rs6897690(GG).

In addition to being the location of IL3, ACSL6, neurogenin 1, SMAD5 and TGFBI, 5q31.1 is also the location for SLC25A48. Liu et al (10) performed a genome-wide association study that identified candidate genes for PD, again in an Ashkenazi Jewish population. They reported that metaanalysis of three datasets supported the association of the SLC25A48 SNP rs4976493(A) with PD. Also, they reported that haplotype analysis observed strongest association for rs4976493-rs4246802(GG). The proband carries rs4976493(AG) and rs4246802(CC).

Again in IHDSF it has been reported that the TAAR6 SNP located at 6q23.2 rs9389011 showed single marker association in schizophrenia whilst haplotype combinations involving the TAAR6 SNPs rs9389011, rs7772821, rs8192622, though not in this combination, gave strongest evidence for association. (11) The proband carries rs9389011(CT), rs7772821(TT) and rs8192622(CC).

SNAP25 is located on chromosome 20p12.2. Fanous et al (12) tested SNAP25 for association with schizophrenia in eighteen SNPs, once again in a sample of IHDSF. It is recalled here that many of the SNP associated with schizophrenia carried by the proband were identified in studies of SNP in IHDSF, suggesting a link between the proband's disorder and IHDSF. Fanous et al found the most significantly associated marker to be rs6039820(CT) risk allele C. They also found eight SNPs to be significantly associated, including rs6032826(AG) risk allele G, rs6077693(AG) risk allele G, rs363043(CT) risk allele C, rs362562(AG) risk allele G, rs6039806(AC) risk allele C, rs6133852(AG) risk allele G. There was significant linkage disquelibrium in the form of five haplotype blocks, two of which were rs6039820/rs362988 and rs6039783/rs363012. The risk allele for rs362988(AG) was the G allele, the risk allele for rs6039783(CT) was the C allele whilst the risk allele for rs363012(AG) was the G allele. They suggested that SNAP25 could be a susceptibility modifier gene in schizophrenia which may infer these SNP are not causative of schizophrenia. The most significantly associated marker was rs6039820 and they noted that this marker was observed to perturb 12 transcription-factor binding sites in in silico analyses. The proband carries SNAP25 SNPs rs6039820(CC), 362988(GG), rs6039783(TT), rs363012(AA), rs6032826(AG), rs6077693(GG), rs363043(CC), rs362562(AG), rs6039806(AC), rs6133852(GG), rs1889189(CT), rs2423487(CT), rs363016(CC) and rs362993(CC).

CSF2RB is inter alia a receptor for IL3. It is located at 22q13.1. The CSF2RB SNPs rs2284031(C/T) and rs909486(C/T) show sex-specific associations with schizophrenia. (13) The proband carries rs2284031(CT) and rs909486(CT).

Sun et al (14) reported finding that the rs6603272(T)-rs6645249(G) haplotype located on the IL3RA gene was significantly associated with familial schizophrenia in a Han Chinese population. IL3RA is located at Xp22.3 or Yp11.3. The proband carries SNPs rs6603272(GT), rs6645249(AG) and rs17883192(CG).

In a genomewide high density SNP linkage study of Japanese families with schizophrenia, rs2048839 (1p21.1) showed significance whilst rs1319956 (14q11.2), rs7149108 (14q12) and rs7988 (20p11.2) showed suggestive evidence of linkage. (15) The proband carries rs2048839(AT), rs1319956(AA), rs7149108(GG) and rs7988(CC).

Finally, in a large-scale study genome wide significant association for schizophrenia were found at 108 locations. (16) Out of the implicated SNP, the proband carries rs11210892(AA), rs11682175(CC), rs6704768(AA), rs17194490(GG), rs7432375(no call), rs10520163(CT), rs1106568(AA), rs1501357(TT), rs3849046(CT), rs10503253(CC), rs4129585(CC), rs7893279(TT), rs9420(GG), rs2693698(AG), rs8042374(AA), rs4702(GG), rs2535627(CT) and rs2007044(AA). Two of the above relate to SNPs (rs2535627 and rs2007044) that are located at CACNA1I which the study noted is located at 22q13.1 which is the same location as the IL3 receptor CSF2RB. Rs2007044 is ranked four in the study which is the highest ranking of any of the above polymorphisms carried by the proband.

Estrogen Response Elements

EREs are specific DNA sequences with high affinity which bind to estrogen receptor and transactivate gene expression in response to estradiol. (16) There is evidence that estrogen signaling may be altered in schizophrenia (17) and that there can be dysregulation of glucocorticoid receptor co-factors in the prefrontal cortex in psychotic illness. (18) BDNF-TRKB which is implicated in schizophrenia pathogenesis can increase transcription at EREs (19) whilst estrogen may regulate BDNF transcription and a putative ERE has been located in the gene encoding BDNF. (20) BDNF blood levels can be reduced in schizophrenia. (21)

L-Arginine:glycine amidinotransferase (AGAT) and the arginine to creatinine pathway.

AGAT is reported to catalyze the production of ornithine and guanidinoacetate, the precursor of creatinine, by transferring an amidino group from arginine to glycine. (22) (23) It is a target of the estrogen receptor and estrogen modulates its expression in chick liver. (24) Sipila et al (25) reported that in gyrate atrophy patients, formation of guanidinoacetate, creatine, and possibly phosphocreatine is inhibited at the transaminidation step by the high concentrations of ornithine.

The proband's plasma ornithine has been repeatedly low, which might suggest that his AGAT activity should be high. However, other test results suggest underactive AGAT: he had low plasma guanidineacetic acid, low urinary creatine, borderline low plasma creatine and undetectable plasma creatinine. Additionally, he has twice had high creatinine excretion. His plasma arginine was repeatedly normal, though when tested by HPLC, Cat-Ion Chromatography Post Column Derivization his plasma arginine was low. His plasma a-N-Acetylarginine was high.

The above is highly suggestive that in the proband there may be abnormalities in AGAT function and/or in the arginine to guandinoacetate to creatine to creatinine pathway.

The proline/polyamine axis.

Ornithine is utilised to produce citrulline which in its turn is utilised to produce arginine. Arginase then acts on arginine to produce urea and ornithine, thus completing the urea cycle.

Amino acid profiles that include abnormally high ornithine have been reported in schizophrenia, (26) but no report has been found of low ornithine occurring. However, low ornithine is a diagnostic tool in P5CS disorders where it would be expected to be accompanied by low proline, arginine and citrulline as well as hyperammonemia. (27)

Whilst the proband's ornithine was repeatedly low his ammonia was twice close to being low (15,20 range 15-50umol/L) and his plasma citrulline was in the normal range.

P5CSlong can be up-regulated by estradiol and P53 and down-regulated by hydrocortisone and dexamethasone. (28) Located close to P5CS at 10q24.3 is CYP17A1 which converts pregnenolone and progesterone to intermediates that are the precursors of estrone and estradiol. (29)

Given that blocks at P5CS result in low proline and low urea cycle intermediates, the existence of the TP53 and estradiol up-regulated/cortisol down-regulated proline pathway might be assumed along with a pathway leading from P5CS to GS to the urea cycle.

Wu et al (30) reported that in cortisol stimulated intestinal polyamine synthesis in neonate suckling pigs proline was a major source of ornithine for this synthesis. Also, that RU486 abolished the stimulatory effect of cortisol on enterocyte OD activity and that cortisol stimulation increased jejunal villus heights. Starvation causes jejunal villus shortening in rats. (31)

In humans this proline to polyamine pathway would be the polyamine pathway.

RU486 is mifepristone which is being considered as a treatment for psychotic depression; it is also reported to decrease the release of IL3. (32) In some circumstances glucocortisoid treatment increases sucrose and maltase activity and decreases lactase activity in piglets, although the magnitude of response is relatively minor compared to amylase and hepatic GOT. (33) (30)

There is evidence to suggest that the proband's proline pathway is overactive whilst his polyamine pathway is underactive. His and his family's symptoms of P5CS/P5CR disorders and his repeat low plasma ornithine indicate underactivity of the proline pathway whilst his marginally high maltase indicates overactivity along the polyamine pathway. His lactase was not low, it was on the low side of normal (2.2 range 1.0-10.0U/g wet wt), but his having been diagnosed with lactose intolerance and his reports that lactose is additionally a trigger, inter alia causing diarrhoea, night terrors and paranoia, are also consistent with his having overactivity along the polyamine pathway. Also, the proband's reports that his disorder is hormonally triggered are consistent with this theory. His amylase and hepatic GOT were not tested, but his sucrose was normal.

Tumour Protein 53 and progranulin

TP53 is a multifunctional protein located at 17p13.1. As stated above, TP53 regulates P5CSlong. It also modulates the expression of PRODH. (34) Thus, it can be presumed that TP53 may regulate, either directly or indirectly, P5CR since P5CR directly reverses the activity of PRODH and since P5CR is downstream from P5CS which is regulated by TP53. TP53 seems to play a role in the regulation of PMP22 (35) (36) and is involved in the metabolism of taurine. (37) (38) TP53 is also implicated in PD and schizophrenia. IL-6 downregulates the expression and activity of p53. (39) IL6 is reported to be elevated in schizophrenia. (40) (41)

The proband additionally carries the TP53 SNP rs2078486(GG) which Yang et al (42) reported may be in linkage disequilibrium with a functional SNP elsewhere which is associated with altered susceptibility to schizophrenia. The proband's taurine was generally low. The proband's IL6 result was only <8pg/mL (interim range <149pg/mL).

Cheung et al (43) report that that p53 wild-type protein nuclei accumulation is associated with GRN protein expression in human hepatocellular carcinoma specimens and that GRN modulates p53 wild-type protein levels in vitro. GRN mutations have been found to be causative of FD and it is suggested that familial cases of dementia associated with GRN mutations may result in schizophrenia. Schizophrenic symptoms may be present for decades before onset of dementia. GRN mutations are also associated with corticobasal syndrome which inter alia causes stiffness.

Memory problems or dementia often precede or follow the psychosis in the proband's family, whilst lower limb stiffness can also occur. In the proband's case, a predilection for sweet foods and mild apathy, disinhibition, loss of empathy, inertia, impaired decision making, metal rigidity, inflexibility and emotional blunting are indicative of schizophrenia associated with FD. (44) (45) His father and other family members had similar symptoms; his father also had intense global sensitivity to cold which is consistent with the hyperalgesia/allodynia which has been reported in GRN deficient mice. (46)

Agmatine and the polyamine/nitric oxide axis

Agmatine is a biogenic amine that is either obtained from diet or synthesised via AD acting on arginine. It is an intermediate on a secondary route to polyamines. It can be transported by hMATE1 (47) which may transport creatinine and nicotinamide. hMATE1 is additionally one of the SMS 17p11.2 deletions. Also, agmatine abnormalities have been reported in schizophrenia. (47)

The proband had elevated urinary creatinine and elevated urinary 1-methyl-5-carboxylamide pyridine (a nicotinamide metabolite) and his plasma creatinine was undetectable suggesting the possibility of hMATE1 being involved in the aetiology of the proband's disorder. He additionally carries rs2270641 (GG) in vehicular monoamine transporter 1 which might be linked to neuropsychiatric disorders. (48) He also reports that severe watery diarrhea and stomach cramps follow his taking agmatine supplements.

Satriano et al (49) reported that catabolism of agmatine to its aldehyde metabolite may act as a gating mechanism at the transition from the iNOS/NO pathway to the arginase/OD/polyamine pathway. He et al (26) discussed evidence that suggested that iNOS competes with arginase for the same substrate, arginine, in producing NO; Kim et al (50) previously proposed that arginase competes for arginine and reduces iNOS activity in genital tissues. In relation to their findings of elevated plasma ornithine in schizophrenia, He et al (26) proposed that substrate arginine consumption and downstream ornithine accumulation in the urea cycle may be induced by aberrant nitric oxide (NO) metabolism that can be found in schizophrenia. If the iNOS/NO pathway is over-expressed it can be speculated that this might result in increased cGMP, given that it appears that iNOS may divert arginine to cGMP via NO. (50) (51) It has also been reported that estrogen treatment of mice upregulates the levels of iNOS mRNA, iNOS protein, and nitric oxide in activated splenocytes. (52)

There is some evidence to support the view that the arginine to iNOS to NO to cGMP pathway in the proband could be over-expressed: it has been noted that the proband's urea was borderline low when he was acutely ill and it is believed the proband's father had low urea, which combined with the proband's low ornithine results suggests the possibility of reduced arginase activity; there is a significant correlation between high carboxyhaemoglobin which the proband had and high cGMP levels; (53) PKG is associated with red blood cell anaemia, which the proband had indications of and which his father had when acutely ill, and with splenomegaly and gastric motility which the father had; if the proband has abnormal cGMP activity, the association between cGMP and diarrhea might provide an explanation for the diarrhea that occurs in this disorder; direct ir infusion of agmatine in rats was reported to produce an increase in creatinine clearance (54) which may explain the proband's high creatinine excretion; also, cGMP, in the brain at least, is reported to be up-regulated by estrogen, (55) which may provide a possible link to the hormonal triggers for the proband's diarrhea; finally, it appears that there is a significant association between agmatine and cGMP levels. (56)

Extending the hypothesis

Retinoic acid increases the expression of TP53. (57) Mutations in retinoic acid induced 1 (RAI1) can result in Smith Magenis Syndrome and RAI1 is regarded as the gene responsible for most phenotypes of Smith Magenis Syndrome. (58) (59) DTNBP1 function appears to be closely related to retinoic acid. (59) Increased expression of RAI1 has been reported in the dorsolateral prefrontal cortex in schizophrenia. (59) Goodman (59) suggested that there could be lowered synthesis of retinoic acid in schizophrenia.

It can be speculated that in the proband's case, retinoic acid may be over-expressed or under-expressed in order to activate TP53 which is required to compensate for the failure of estradiol to activate the proline pathway. A protective DTNBP1 haplotype carried by the proband may play a role in favourably altering the expression of retinoic acid. However, the altered expression of retinoic acid then results in abnormal RAI1 activity and thus the symptoms of Smith Magenis Syndrome that are comorbid with psychosis in the proband's family. Inconsistent with this proposal is the fact that the phenotype consequences of RAI1 overexpression in mice are not consistent with the symptoms present in the proband's family. (59)

Further evidence in support of the hypothesis

Most of the hypothesis as regards the metabolic derangement at play in the proband was developed before his genetic tests indicated that 5q31.1 was heavily implicated in his disorder. Below is detailed some of the evidence that suggests 5q31.1 could play a central role in controlling the metabolic pathways that the hypothesis suggests are involved in the disorder. Of course, this might be coincidental.

Schneider et al (60) reported that homogeneous interleukin 3 enhances arginase activity in murine hematopoietic cells which suggests a role for 5q31.1 in the control of whether arginine is utilised by the arginine/arginase/OD/polyamine pathway and/or the arginine/iNOS/NO pathway.

Anderson's disease, also known as chylomycron retention disease, is caused by mutations in the SAR1B gene which is located at 5q31.1. The proband and his father both suffered from anaemia/indications of anaemia as well as diarrhea and the proband has suffered from peripheral neuropathy, all of which are symptoms of Anderson's disease. Further, the proband and his father both had lipid abnormalities, though their abnormalities were not identical to the lipid abnormalities that present in Anderson's disease.

Zhu et al (61) reported that the soluble guanylate cyclase signal transduction pathway is modulated by hydroxyurea regulated y-globin expression which also increases NO production and also reported that SAR1 significantly increases y-globin expression. Soluble guanylate cyclase plays a key role in cGMP metabolism. This might indicate that a connection between SAR1 and cGMP may account for the diarrhea that presents in Anderson's disease.

PDE6A which is cGMP specific is located close to 5q31.1 at 5q31.2 which also suggests a role for this location in cGMP metabolism.

Prolyl 4-hydroxylase which is located at 5q31.1 and which is downstream of proline catalyzes the formation of 4-hydroxyproline, (62) suggesting the possibility that 5q31.1 may be involved in the control of the proline pathway. Prolyl 4-

hydroxylase has also been reported to show partial homology to the estrogen-binding domain of the estrogen receptor. (63)

Pochini et al (64) reported that OCTN1 (SLC22A4), which is located at 5q31.1, when reconstituted in liposomes catalyzes acetylcholine transport which is defective in the mutant L503F associated with Crohn's disease and that the acetylcholine transport was almost completely inhibited by the polyamines spermine and spermidine at concentrations as low as 0.5 mM. It would be interesting to know if OCTN1 activity plays any role in regulating polyamine synthesis because this would support 5q31.1 playing a role in the regulation of the polyamine pathway. It would also be interesting to know if OCTN1 plays a role in creatinine transport.

DISCUSSION

Regrettably, it has not been possible to provide the clinical history of all family members nor to carry out genetic and biomarker tests on other family members. As a result the proposed hypothesis remains very provisional.

Even so, the weight of evidence outlined above does suggest that this psychotic disorder probably has a distinct and identifiable phenotype. The likely significance of the P5CR/P5CS symptoms and conditions in the proband and his family is supported by the proband's abnormal test results. The association of the disorder with 17p11.2 disorders is not as clear, but may prove significant.

In light of the consistent presentation of the disorder over several generations, it is more likely that a common genetic cause is at play, though one which is still influenced by other genetic and environmental factors. Haplotype inheritance of SNPs at 5q31.1 appears to be strongly implicated.

The proband's paternal family is from Hawarden on the Welsh Borders. It is reasonable to assume there may have been a family connection to Ireland and thus to IHDSF. Over-expression of the polyamine pathway and a consequential increase in jejunal villus heights may have provided a survival advantage in a starving population such as the Irish during the potato famine.

It may be speculated that schizophrenic patients with high ornithine and high IL6 may be suffering from a genetic defect that results in malfunctioning of the cortisol regulation of the proline and polyamine pathways.

It is concluded that the proband could have under-expression of an estradiol up-regulated proline pathway along with consequential over-expression of a cortisol up-regulated polyamine pathway. Also, he does carry SNPs that are significantly associated with high density schizophrenia which are located in and around ERE which transactivate gene expression in response to estradiol.

However, whilst it may be the case that there is an as yet incompletely elucidated connection between 5q31.1 and the metabolic pathways that the hypothesis suggests are malfunctioning in the proband, it remains a leap to suggest that defective ERE function is a significant or causative factor in his disorder.

Alternative hypotheses are that:

1. The SNPs located at IL3/ACSL6 at 5q31.1 result in abnormal IL3 activity which in turn causes abnormal arginase activity which results in the same up-regulated and down-regulated pathways mentioned above; or
2. SNPs that are linked to various genes at 5q31.1 result in over-expression and under-expression of the pathways mentioned above perhaps as a result of some of these SNPs causing under or over-function of EREs at this location.

It is thus important to establish whether or not P5CS and/or the proline and/or polyamine pathways are regulatory targets of ERE, particularly ERE1/2/3/4/6 located at 5q31.1. If this were established it would bring to the fore the question of what role HRE may play in other disorders that can result in schizophrenia. Of particular interest would be any HRE located in or around other SNP that have significant association with schizophrenia.

If the SNPs at 5q31.1 that are associated with schizophrenia play a central role in this familial disorder and if they result in under-function of the proline pathway, it may be expected that evolutionary pressure may have resulted in particular patterns of SNPs being carried at other key locations such as TP53, PRODH and P5CR. Further work is indicated to establish if this is the case.

Declaration

The writer has no formal medical training and is not a professional researcher.

Conflict of interest:

Competing interest - none declared.

Ethics committee approval

This was not required because the proband is the writer.

Figure 1.

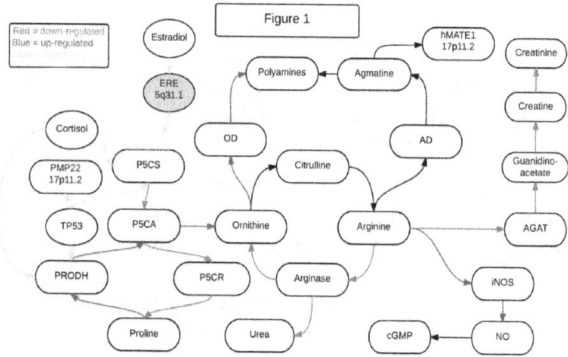

SUPPLEMENTAL MATERIAL.

A – The proband's abnormal results;

B – Lists of the proband's and his family members' symptoms and conditions.

A: THE PROBAND'S ABNORMAL RESULTS

The proband had the following abnormal results:

1. Plasma ornithine - 46,46 48,48,49,55,60,62 range 80-150aemol/L;
2. Plasma taurine - 40,44,45,45,47,48,50,59 range 50-100aemol/L;

3. Plasma serine - 57,63,65,69,72,86,98 range 90-200aemol/L;
4. Plasma arginine 0.163 range 110+/-23.9umol/L;
5. Plasma arginine 75, 94, 84, 67, 87, 86, 100, 100 range 34-118umol/L;
6. Plasma arginine 71 range 40-200aemol/L;
7. Plasma homoarginine ND range 1.98+/-0.634umol/L;
8. Plasma guandinosuccinic acid 0.038 range 0.259 +/-0.096umol/L;
9. Plasma guanidinoacetic acid 0.274 range 2.61+/-0.517umol/L;
10. Plasma beta-guandinopropionic acid was detectable at 69umol/L;
11. Plasma creatine 20 range 30.1+/-12.3umol/L;
12. Plasma creatinine ND range 80.8+/- 17.7;
13. Creatinine excretion - 18,19.8 range 5-16mmol/day;
14. Urine creatinine conc. 7.8mmol/L;
15. Urine creatine 3.1 range 3.4-191mmol/mol creatinine;
16. Urine creatinine 4457.5 umol/L;
17. Urine guanidine ND range 1.1+/-0.38mmol/mol creatinine;
18. Urine methylguanidine ND range 0.3+/-0.14mmol/mol creatinine;
19. Urea - 3.0 range 3.0-8.5mmol/L;
20. Urine arginine/creatinine ratio 2, 2, 4, 1, 1, 2, range 0-16umol/mmol.
21. Urine arginine 1.0 range 2.4+/-1.79 mmol/mol creatinine;
22. Urine a-N-Acetylarginine 5.8 range 2.5+/-1.08;
23. Urine guanidinosuccinic acid 5.8 range 2.8+/-1.02mmol/mol creatinine;

24. Urine glutamic acid/creatinine ratio - 2,3,4,4,6,7 range 1-5umol/mmol;
25. Urine 1-methyl-5-carboxylamide pyridine - 144 range 25-110umol/d;
26. RBC 4.3 - range (4.5-6.5)10*12/L;
27. Iron saturation 47% range 10-45;
28. Trans sat 23% range 15-50%;
29. IL6 <8 interim range <149pg/mL;
30. Haemoglobin – 131 range 130-180g/L;
31. Haematocrit - 0.38 range 0.40-0.54;
32. Carboxyhaemoglobin - 3.1 range 0.00-1.5%;
33. Hdl - 0.90 range 1-3mmol/L;
34. Hdl 1.1 range>1.0mmol/L;
35. Chol/hdl - 5.3 range 0-4.5mmol/L;
36. LDL - 3.6 range 0.5-3.5mmol/L;
37. LDL 3 range 0-4mmol/l;
38. Maltase - 786 range 79-655U/g prot:

The ranges are usually those provided by the labs that performed the tests. Where there have been abnormal and normal results the normal results are also provided. Where labs have performed multiple tests the different results are separated by commas.

B: LISTS OF THE PROBAND'S AND HIS FAMILY MEMBERS' SYMPTOMS AND CONDITIONS

The psychotic/schizophrenic symptoms found in at least three generations of the proband's family.

The psychotic/schizophrenic symptoms suffered by the proband include:

1. Jealous and then persecutory delusional beliefs with episodes becoming of increasing frequency, duration and severity;
2. Insomnia;
3. Night terrors;
4. Social withdrawal;
5. Emotional blunting;
6. Loss of ability to function at work resulting in loss of employment; and
7. Obsessiveness.

The psychotic/schizophrenic symptoms suffered by the proband's father include:

1. Jealous and then persecutory delusional beliefs with episodes becoming of increasing frequency, duration and severity;
2. Insomnia;
3. Night terrors;
4. Loss of ability to function at work resulting in early retirement; and
5. Obsessiveness.

The psychotic/schizophrenic symptoms suffered by the paternal aunt include:

1. Severe, frequent and often florid persecutory delusions with episodes becoming of increasing frequency, duration and severity;
2. Early loss of ability to function at work resulting in blighted employment history;
3. Severe social withdrawal; and
4. Obsessiveness.

The psychotic/schizophrenic symptoms suffered by the paternal grandmother include:

1. Persecutory delusions with episodes becoming of increasing frequency, duration and severity; and
2. Obsessiveness.

The symptoms/conditions associated with Smith Magenis Syndrome, PMP22 peripheral neuropathy with liability to pressure palsies and 17p11.2 conditions found in at least three generations of the proband's family.

The 17p11.2 disorder symptoms/conditions suffered by the proband include:

1. Diarrhoea (he had monthly diarrhea, fever and lethargy/stupor from one month old to about nine years old which then improved but did not dissipate completely before suffering deteriorating diarrhea in early middle age);
2. Constipation;
3. Irritability;
4. Glue ear and vulnerability to ear infections;
5. Prognathism;
6. Peripheral neuropathy;
7. Pressure palsies;
8. Myopia;
9. Irritability;
10. Temper tantrums as a child;
11. Daytime sleepiness and night-time insomnia;
12. Picking at nails and cuticles including occasionally picking away the base of the nail;
13. Having an occasional mild rocking sensation; and
14. Vulnerability to hoarse voice.

The 17p11.2 symptoms/conditions suffered by the proband's father include:

1. Diarrhoea;
2. Irritability;
3. Prognathism;
4. Myopia;
5. Irritability;
6. Daytime sleepiness and night-time insomnia; and
7. Kidney cysts.

The 17p11.2 symptoms/conditions suffered by the paternal aunt include:

1. Prognathism;
2. Irritability;
3. Severe hoarse voice; and
4. Symptoms of peripheral neuropathy.

The 17p11.2 symptoms/conditions suffered by the paternal grandmother include:

1. Prognathism; and
2. Daytime sleepiness and night-time insomnia.

The dementia and Parkinson's symptoms and conditions found in three generations of the proband's family.

The early onset symptoms of dementia and Parkinson's disease suffered by the proband include:

1. Early onset short term memory problems;
2. Early onset memory black outs;
3. Impaired verbal memory;
4. Adverse reaction to antipsychotics
5. Constipation;
6. Hyposmia;
7. Impaired concentration;
8. Orthostatic symptoms;
9. Anxiety;
10. Delusions;
11. Vivid dream imagery;
12. Poor multitasking;
13. Fluctuating concentration and attention;
14. Tendency to zone out;
15. Rigidity and stiffness;
16. Decreased change of facial expression;
17. Very mild drooling from corners of mouth;
18. Apathy;
19. Daytime sleepiness;
20. Insomnia;
21. Dizziness; and
22. High frequency urinating.

The early onset symptoms of dementia and Parkinson's disease suffered by the proband's father include:

1. Short term memory problems;
2. Memory black outs;
3. Impaired concentration;
4. Hand tremor;

5. Delusions;
6. Vivid dream imagery;
7. Poor multitasking;
8. Daytime sleepiness;
9. Insomnia; and
10. High frequency urinating;

The symptoms of dementia and Parkinson's disease, if any, suffered by the paternal aunt are unknown. She had suffered from stiffness that preceded rheumatoid arthritis.

The paternal grandmother had diagnosed dementia associated with delusional psychosis and Parkinson's disease.
The proband's maternal grandmother suffered from late-onset Alzheimer's disease.

The P5CS/P5CR disorder symptoms and conditions found in at least three generations of the proband's family.

The P5CS/P5CR disorder symptoms/conditions suffered by the proband include:

1. Worsening prognathism;
2. Wide flat feet and bilateral bunionated big toes;
3. Peripheral neuropathy;
4. Indications of lax skin on the back of the hands;
5. Memory problems.

The P5CS/P5CR disorders symptoms/conditions suffered by the proband's father include:

1. Worsening prognathism;
2. Joint laxity;
3. Bow legs;

4. Hypertension;
5. Hiatal hernia;
6. Diverticulitis;
7. Bilateral cataracts;
8. Memory problems;
9. Atherosclerosis;
10. Splenomegaly;
11. Lax skin on the back of the hand; and
12. His cause of death which was haemorrhaging of hiatal hernia and oesophageal varices brought on by portal hypertension is consistent with cutis laxa.

The P5CS/P5CR disorder symptoms and conditions suffered by the paternal aunt include:

1. Worsening prognathism; and
2. Symptoms of peripheral neuropathy.

The P5CS/P5CR disorder symptoms and conditions suffered by the paternal grandmother include:

1. Prognathism; and
2. Memory problems.

Other

Symptoms which may be significant found in at least two of those family members who have suffered significant symptoms of the disorder but which do not fit into the categories listed above include: dislike of bright light and avoidance of bright sunlight; vulnerability to heat stroke; sensory noise intolerance;

growing excessively long nails; sores on the back of the fingers; weakness on waking; weak grip; white mouth ulcers; delayed reaction/processing time and consequential mild discoordination; and dyslexia with homonyms.

Note

The proband's paternal grandmother had two children – the proband's father and aunt. The proband has three siblings and the proband and each of his siblings each have two children.

One of the proband's siblings reports he and his children display no symptoms of P5CR/P5CS disorders, Smith Magenis Syndrome, psychosis, Parkinson's disease and dementia. The proband's other two siblings have not co-operated to provide details of any such symptoms/conditions in themselves and their children and the proband's reports in this regard are not recorded here.

The proband's grandmother, father and aunt are all deceased and the proband's knowledge of their medical histories is limited which consequently means that the list of symptoms/conditions suffered by them is not comprehensive.

References

1. Smith A, Gropman A , Bailey-Wilson J, Goker-Alpan O, Elsea S, Blancato J, et al. Hypercholesterolemia in children with Smith-Magenis syndrome.: del (17)(p11.2p11.2) Genetics in Medicine (2002) 4, 118–125; doi:10.1097/00125817-20020500.

2. Chen X, Wang X, Hossain S, O'Neil FA, Walsh D, van den Oord E, et al. Interleukin 3 and schizophrenia: the impact of sex and family history. Mol Psychiatry. 2007 Mar;12(3):273-82. Epub 2006 Dec 19.

3. Edwards TL, Wang X, Chen Q, Wormly B, Riley B, , O'Neill FA, Walsh D, Ritchie MD, Kendler KS, Chen X. Interaction between interleukin 3 and dystrobrevin-binding protein 1 in schizophrenia. Schizophr Res. 2008. Dec.; 106(2-3):208-17. doi: 10.1016/j.schres.2008.

4. Guo AY, Sun J, Riley BP, Thiselton DL, Kendler KS, Zhao Z. The dystrobrevin-binding protein 1 gene: features and networks. Mol Psychiatry. 2009 Jan;14(1):18-29. doi: 10.1038/mp.2008.88. Epub 2008 Jul 29.

    Chen X, Wang X, O'Neill AF, Walsh D, Kendler KS. Variants in the catechol-o-methyltransferase (COMT) gene are associated with schizophrenia in Irish high-density families. Mol Psychiatry. 2004 Oct;9(10):962-7.

6. Shifman S, Bronstein M, Sternfeld M , Pisanté-Shalom A, Lev-Lehman E, Weizman A , Reznik I, Spivak B, Grisaru N, Karp L, Schiffer R , Kotler M, Strous RD, Swartz-Vanetik M, Knobler HY, Shinar E, Beckmann JS, Yakir B, Risch N, Zak NB , Darvasi

A. A highly significant association between a COMT haplotype and schizophrenia. Am J Hum Genet. 2002 Dec;71(6):1296-302. Epub 2002 Oct 25. .

7. Sweet RA, Devlin B, Pollock BG, Sukonick DL, Kastango KB, Bacanu SA, Chowdari KV, DeKosky ST, Ferrell RE. Catechol-O-methyltransferase haplotypes are associated with psychosis in Alzheimer disease. Mol Psychiatry. 2005 Nov;10(11):1026-36. .

8. Fanous AH, Chen X, Wang X, Amdur RL, O'Neill FA, Walsh D, Kendler KS. Association between the 5q31.1 gene neurogenin1 and schizophrenia. Am J Med Genet B Neuropsychiatr Genet. 2007 Mar 5;144B(2):207-14.

9. Zaharieva I, Georgieva L, Nikolov I, Kirov G, Owen MJ, O'Donovan MC, Toncheva D. Association study in the 5q31-32 linkage region for schizophrenia using pooled DNA genotyping. 2008 Feb 25;8:11. doi: 10.1186/1471-244X-8-11.

10. Liu X, Cheng R, Verbitsky M, Kisselev S, Browne A , Mejia-Sanatana H, Louis ED, Cote LJ, Andrews H, Waters C, Ford B, Frucht S, Fahn S , Marder K, Clark LN, Lee JH. Genome-wide association study identifies candidate genes for PD in an Ashkenazi Jewish population. BMC Med Genet. 2011 Aug 3;12:104. doi: 10.1186/1471-2350-12-104.

11.
Vladimirov V, Thiselton DL, Kuo PH, McClay J, Fanous A, Wormley B, Vittum J, Ribble R, Moher B, van den Oord E, O'Neill FA, Walsh D, Kendler KS, Riley BP. A region of 35 kb containing the trace amine associate receptor 6 (TAAR6) gene is associated with schizophrenia in the Irish study of

high-density schizophrenia families. Mol Psychiatry. 2007 Sep;12(9):842-53. Epub 2007 May 15.

12. Fanous AH, Zhao Z, van den Oord EJ, Maher BS, Thiselton DL, Bergen SE, Wormley B, Bigdeli T, Amdur RL, O'Neill FA, Walsh D, Kendler KS, Riley BP. Association study of SNAP25 and schizophrenia in Irish family and case-control samples. AMJ Med Genet B Neuropsychiatr Genet. 2010 Mar 5;153B(2):663-74. doi: 10.1002/ajmg.b.31037.

Chen Q, Wang X, O'Neill FA, Walsh D, Fanous A, Kendler KS, Chen X. Association study of CSF2RB with schizophrenia in Irish family and case - control samples. Molecular psychiatry. 2008 Oct;13(10):930-8. Epub 2007 Jul 31.

14. Sun S, , Wei J, Li H, Jin S, Li P, Ju G, Liu Y, Zhang XY. A family-based study of the IL3RA gene on susceptibility to schizophrenia in a Chinese Han population. Brain Res. 2009 May 1;1268:13-6. doi: 10.1016/j.brainres.2009.02.071. Epub 2009 Mar 10. .

Arinami et al. Genomewide high-density SNP linkage analysis of 236 Japanese families supports the existence of schizophrenia susceptibility loci on chromosomes 1p, 14q, and 20p. Am J Hum Genet. 2005 Dec;77(6):937-44. Epub 2005 Oct 12.

16. Schizophrenia Working Group of the Psychiatric Genetics Consortium. Biological insights from 108 schizophrenia-associated genetic loci. Nature. 2014 Jul 24;511(7510):421-7. doi: 10.1038/nature13595. Epub 2014 Jul 22.

16. Klinge CM. Estrogen receptor interaction with estrogen

response elements.: Oxford Journals. Science & Mathematics. Nucleic Acids Research Volume 29, Issue 14. Pp. 2905-2919.

17. Wong J, Weickert CS. Transcriptional interaction of an estrogen receptor splice variant and ErbB4 suggests convergence in gene susceptibility pathways in schizophrenia.: Biol Chem. 2009 Jul 10;284(28):18824-32. doi: 10.1074/jbc.M109.013243. Epub 2009 May 13.

18. Sinclair D, Fillman SG, Webster MJ, Weickert CS. Dysregulation of glucocorticoid receptor co-factors FKBP5, BAG1 and PTGES3 in prefrontal cortex in psychotic illness. Sci Rep. 2013 Dec 18;3:3539. doi: 10.1038/srep03539.

19. Wong J, Woon HG, Weickert CS. Full length TrkB potentiates estrogen receptor alpha mediated transcription suggesting convergence of susceptibility pathways in schizophrenia. Mol Cell Neurosci. 2011 Jan;46(1):67-78. doi: 10.1016/j.mcn.2010.08.007. Epub 2.

20. Sohrabji F, Miranda RC, Toran-Allerand CD. Identification of a putative estrogen response element in the gene encoding brain-derived neurotrophic factor. Proc Natl Acad Sci U S A. 1995 Nov 21;92(24):11110-4.

21. Green MJ, Matheson SL, Shepherd A, Weickert CS, Carr VJ. Brain-derived neurotrophic factor levels in schizophrenia: a systematic review with meta-analysis. Mol Psychiatry. 2011 Sep;16(9):960-72. doi: 10.1038/mp.2010.88. Epub 2010 Aug 24.

22. Fritsche E, Humm A, Huber R. The Ligand-induced Structural

Changes of Humanl-Arginine:Glycine Amidinotransferase. A MUTATIONAL AND CRYSTALLOGRAPHIC STUDY. J Biol Chem. 1999 Jan 29;274(5):3026-32.

23. Humm A, Fritsche E, Steinbacher S, Huber R. Crystal structure and mechanism of human arginine:glycine amidinotransferase: a mitochondrial enzyme involved in creatine biosynthesis. EMBO J. 1997 Jun 16;16(12):3373-85.

24. Zhu Y, Evans Ml. Estrogen modulates the expression of L-arginine:glycine amidinotransferase in chick liver. Mol Cell Biochem. 2001 May;221(1-2):139-45.

25. Sipila I, Simell O, Arjomaa P. Gyrate atrophy of the choroid and retina with hyperornithinemia. Deficient formation of guanidinoacetic acid from arginine. J Clin Invest. 1980 Oct;66(4):684-7.

26. He, Yu, Giegling, Xie, Hartmann, Prehn, Adamski, Kahn, Li, Illig, Wang-Sattler and Rujescu. Schizophrenia shows a unique metabolic signature in plasma. Translational Psychiatry. 2012 Aug 14;2:e149. doi: 10.1038/tp.2012.76.

27. Baumgartner, Matthias, Hu, Chien-an A, Shlomo. Hyperammonemia with reduced ornithine, citrulline, arginine and proline: a new inborn error caused by a mutation in the gene encoding

28. Hu, Khalil, Zhaorigetu, Liu, Typer, Wan and Valle. Human Delta 1-pyrroline-5-carboxylate synthase function and regulation. Amino Acids. Nov 2008; 35(4): 665-672.

29. Olson SH, Orlow I, Bayuga S, Sima C , Bandera EV, Pulick K,

Faulkner S, Tommasi D, Egan D, Roy P, Wilcox H, Asya A, Modica I, Asad H, Soslow R, Zauber AG. Variants in hormone biosynthesis genes and risk of endometrial cancer. Cancer causes control. 2008 Nov;19(9):955-63. doi: 10.1007/s10552-008-9160-7. Epub 2008 Apr 25.

30. Wu G, Flynn NE, Knabe DA. Enhanced intestinal synthesis of polyamines from proline in cortisol-treated piglets. Am J Physiol Endocrinol Metab 279: E395–E402, 2000.

31. Buts JP, Vijverman V, Barudi C, De Keyser N, Maldague P, Dive C. Refeeding after starvation in the rat: comparative effects of lipids, proteins and carbohydrates on jejunal and ileal mucosal adaptation. Eur J Clin Invest. 1990 Aug;20(4):441-52.

32. Antonakis N, Georgoulias V, Margioris AN, Stournaras C, Gravanis A. In vitro differential effects of the antiglucocorticoid RU486 on the release of lymphokines from mitogen-activated normal human lymphocytes. J Steroid Biochem Mol Biol. 1994 Oct;51(1-2).

33. Chapple RP, Cuaron JA, Easter RA. Effect of glucocorticoids and limiting nursing on the carbohydrate digestive capacity and growth rate of piglets. J Anim Sci. 1989 Nov;67(11):2956-73.

34. Raimondi I, Ciribilli Y, Monti P, Bisio A, Pollegioni L, Fronza G, Inga A, Campomenosi. P53 family members modulate the expression of PRODH, but not PRODH2, via intronic p53 response elements.0. PLoS One. 2013 Jul 8;8(7):e69152. doi: 10.1371/journal.pone.0069152. Print 2013.

35. Attardi LD, Reczek EE, Cosmas C, Demicco EG, McCurrach ME, Lowe SW, Jacks T. PERP, an apoptosis-associated target of p53, is a novel member of the PMP-22/gas3 family. Genes Dev. 2000 Mar 15;14(6):704-18.

36. Mallette FA, Calabrese V, Ilangumaran S, Ferbeyre G. SOCS1, a novel interaction partner of p53 controlling oncogene-induced senescence. Aging (Albany NY). 2010 Jul;2(7):445-52.

37. Han X, Patters AB, Chesney RW. Transcriptional repression of taurine transporter gene (TauT) by p53 in renal cells. J Biol Chem. 2002 Oct 18;277(42):39266-73. Epub 2002 Aug 5.

38. Zhang EB, Yin DD, Sun M, Kong R, Liu XH, You LH, Han L, Xia R, Wang KM, Yang JS, De W, Shu YQ, Wang ZX. P53-regulated long non-coding RNA TUG1 affects cell proliferation in human non-small cell lung cancer, partly through epigenetically regulating HOXB7 expression. Cell Death Dis. 2014 May 22;5:e1243. doi: 10.1038/cddis.2014.201.

39. Brighenti E, Calabrese C, Liguori FA, Giannone FA, Trere D, Montanaro L, et al. Interleukin 6 downregulates p53 expression and activity by stimulating ribosome biogenesis: a new pathway connecting inflammation to cancer. Oncogene. 2014 Aug 28;33(35):4396-406. doi: 10.1038/onc.2014.1. Epub 2014 Feb 17.

40. Naudin J, Mege JL, Azorin JM, Dassa D. Elevated circulating levels of IL-6 in schizophrenia. Schizophr Res. 1996 Jul 5;20(3):269-73.

41. Naudin J, Capo C, Giusano B, Mege JL, Azorin JM. A differential role for interleukin-6 and tumor necrosis factor-

alpha in schizophrenia? Schizophr Res. 1997 Aug 29;26(2-3):227-33.

42. Yang Y, Xiao Z, Chen W, Sang H, Guan Y, Peng Y, Zhang D, Gu Z, Qian M, He G , Qin W, Li D, Gu N, He L. Tumor suppressor gene TP53 is genetically associated with schizophrenia in the Chinese population. Neurosci Lett. 2004 Oct 14;369(2):126-31.

43. Cheung ST, Wong SY, Lee YT, Fan ST. GEP associates with wild-type p53 in hepatocellular carcinoma. Oncol Rep. 2006 Jun;15(6):1507-11.

44. Cooper JJ, Ovsiew F. The relationship between schizophrenia and FD. J Geriatr Psychiatry Neurol. 2013 Sep;26(3):131-7. doi: 10.1177/0891988713490992. Epub 2013 Jun 3.

45. Leyton CE, Hodges JR. Frontotemporal dementias: Recent advances and current controversies. Ann Indian Acad Nerol. 2010 Dec;13(Suppl 2):S74-80. doi: 10.4103/0972-2327.74249.

46. Albuquerque B, Häussler A, Vannoni E, Wolfer DP, Tegeder I. Learning and memory with neuropathic pain: impact of old age and progranulin deficiency. Front Behav Neurosci. 2013 Nov 22;7:174. doi: 10.3389/fnbeh.2013.00174. eCollection 2013.

47. Winter TN, Elmquist WF, Fairbanks CA. OCT2 and MATE1 provide bidirectional agmatine transport. Mol Pharm. 2011 Feb 7;8(1):133-42. doi: 10.1021/mp100180a. Epub 2010 Dec 3; or

Uzbay T, Goktalay G, Kayir H, Eker SS, Sarandol A, Oral S, Buyukuysal L, Ulusoy G, Kirli S. Increased plasma agmatine levels in patients with schizophrenia. J Psychiatr Res. 2013 Aug;47(8):1054-60. doi: 10.1016/j.jpsychires.2013.04.004. Epub 2013 May 7.

48. Lohoff FW. Genetic variants in the vesicular monoamine transporter 1 (VMAT1/SLC18A1) and neuropsychiatric disorders. Methods Mol Biol. 2010;637:165-80. doi: 10.1007/978-1-60761-700-6_9.

49. Satriano J. Agmatine: at the crossroads of the arginine pathways. Ann N Y Acad Sci. 2003 Dec;1009:34-43.

50. Kim NN, Christianson DW, Traish AM. Role of arginase in the male and female sexual arousal response. J Nutr. 2004 Oct;134(10 Suppl):2873S-2879S; discussion 2895S.

51. Rapôso C, Luna RL, Nunes AK, Thomé R, Peixoto CA. Role of iNOS-NO-cGMP signaling in modulation of inflammatory and myelination processes. Brain Res Bull. 2014 May;104:60-73. doi: 10.1016/j.brainresbull.2014.04.002. Epub 2014 Apr 13.

52. Karpuzoglu E, Ahmed SA. Estrogen regulation of nitric oxide and inducible nitric oxide synthase (iNOS) in immune cells: implications for immunity, autoimmune diseases, and apoptosis. Nitric Oxide. 2006 Nov;15(3):177-86. Epub 2006 May 2.

53. Van Bel F, Latour V, Vreman HJ, Wong RJ, Stevenson DK, Steendijk P, Egberts J, Krediet TG. Is carbon monoxide-mediated cyclic guanosine monophosphate production responsible for low blood pressure in neonatal respiratory distress syndrome? J Appl Physiol (1985). 2005 Mar;98(3):1044-9. Epub 2004 Oct 29.

54. Penner SB, Smyth DD. Natriuresis following central and peripheral administration of agmatine in the rat. Pharmacology. 1996 Sep;53(3):160-9.

55. Palmon SC, Williams MJ, Littleton-Kearney MT, Traystman RJ, Kosk-Kosicka D, Hurn PD. Estrogen increases cGMP in selected brain regions and in cerebral microvessels. J Cereb Blood Flow Metab. 1998 Nov;18(11):1248-52.

56. Schwartz D, Peterson OW, Mendonca M, Satriano J, Lortie M, Blantz RC. Agmatine affects glomerular filtration via a nitric oxide synthase-dependent mechanism. Am J Physiol. 1997 May;272(5 Pt 2):F597-601.

57. Mrass P, Rendl M, Midner M, Gruber F, Lengauer B, Ballaun C, et al. Retinoic acid increases the expression of p53 and proapoptotic caspases and sensitizes keratinocytes to apoptosis: a possible explanation for tumor preventive action of retinoids. Cancer Res. 2004 Sep 15;64(18):6542-8.

58. Slager RE, Newton TL, Vlangos CN, Finucane B, Elsea SH. Mutations in RAI1 associated with Smith-Magenis syndrome. Nat Genet. 2003 Apr;33(4):466-8. Epub 2003 Mar 24.

59. Girirajan S, Vlangos CN, Szomju BB, Edelman E, Trevors CD, Dupuis L, et al. Genotype-phenotype correlation in Smith-Magenis syndrome: evidence that multiple genes in 17p11.2 contribute to the clinical spectrum. Genet Med. 2006 Jul;8(7):417-27.

Guo AY, Sun J, Riley BP, Thiselton DL, Kendler KS, Zhao Z. The dystrobrevin-binding protein 1 gene: features and networks. Mol Psychiatry. 2009 Jan;14(1):18-29. doi:

10.1038/mp.2008.88. Epub 2008 Jul 29..

Haybaeck J, Postruznik M, Miller CL, Dulay JR, Llenos IC, Weis S. Increased expression of retinoic acid-induced gene 1 in the dorsolateral prefrontal cortex in schizophrenia, bipolar disorder, and major depression. Neuropsychiatr Dis Treat. 2015 Feb 4;11:279-89. doi: 10.2147/NDT.S72536. eCollection 2015.

Goodman AB. Microarray results suggest altered transport and lowered synthesis of retinoic acid in schizophrenia. Mol Psychiatry. 2005 Jul;10(7):620-1.

Haybaeck J, Postruznik M, Miller CL, Dulay JR, Llenos IC, Weis S. Increased expression of retinoic acid-induced gene 1 in the dorsolateral prefrontal cortex in schizophrenia, bipolar disorder, and major depression. Neuropsychiatr Dis Treat. 2015 Feb 4;11:279-89. doi: 10.2147/NDT.S72536. eCollection 2015.

60. Schneider E, Ihle JN, Dy M. Homogeneous interleukin 3 enhances arginase activity in murine hematopoietic cells. Lymphokine Res. 1985 Spring;4(2):95-102.

61. Zhu J, Chin K, Aerbajinai W, Kumkhaek C, Li H, Rodgers GP. Hydroxyurea-inducible SAR1 gene acts through the Giα/JNK/Jun pathway to regulate γ-globin expression. Blood. 2014 Aug 14;124(7):1146-56. doi: 10.1182/blood-2013-10-534842. Epub 2014 Jun 9.

62. Kukkola L, Hieta R, Kivirikko KI, Myllyharju J. Identification and characterization of a third human, rat, and mouse collagen prolyl 4-hydroxylase isoenzyme. J Biol Chem. 2003 Nov 28;278(48):47685-93. Epub 2003 Sep 18.

63. Kivirikko KI, Myllylä R, Pihlajaniemi T. Protein hydroxylation: prolyl 4-hydroxylase, an enzyme with four cosubstrates and a multifunctional subunit. FASEB J. 1989 Mar;3(5):1609-17.

64. Pochini L, Scalise M, Galluccio M, Pani G , Siminovitch KA, Indiveri C. The human OCTN1 (SLC22A4) reconstituted in liposomes catalyzes acetylcholine transport which is defective in the mutant L503F associated to the Crohn's disease. Biochim Biophys Acta.2012 Mar;1818(3):559-65. doi: 10.1016/j.bbamem.2011.12.014. Epub 2011 Dec 21.

www.ingramcontent.com/pod-product-compliance
Lightning Source LLC
Chambersburg PA
CBHW072259170526
45158CB00003BA/1116